GOOD LABORATORY PRACTICE IN THE TESTING OF CHEMICALS

FINAL REPORT OF THE GROUP OF EXPERTS
ON GOOD LABORATORY PRACTICE

ORGANISATION FOR ECONOMIC CO-OPERATION AND DEVELOPMENT

The Organisation for Economic Co-operation and Development (OECD) was set up under a Convention signed in Paris on 14th December 1960, which provides that the OECD shall promote policies designed:
- to achieve the highest sustainable economic growth and employment and a rising standard of living in Member countries, while maintaining financial stability, and thus to contribute to the development of the world economy;
- to contribute to sound economic expansion in Member as well as non-member countries in the process of economic development;
- to contribute to the expansion of world trade on a multilateral, non-discriminatory basis in accordance with international obligations.

The Members of OECD are Australia, Austria, Belgium, Canada, Denmark, Finland, France, the Federal Republic of Germany, Greece, Iceland, Ireland, Italy, Japan, Luxembourg, the Netherlands, New Zealand, Norway, Portugal, Spain, Sweden, Switzerland, Turkey, the United Kingdom and the United States.

Publié en français sous le titre :

LES BONNES PRATIQUES DE LABORATOIRE
DANS LES ESSAIS DES PRODUITS CHIMIQUES

© OECD, 1982
Application for permission to reproduce or translate all or part of this publication should be made to:
Director of Information, OECD
2, rue André-Pascal, 75775 PARIS CEDEX 16, France.

FOREWORD

Chemicals control laws passed in OECD Member countries in the 1970s and 1980s generally call for some testing and assessing of chemicals. It is therefore of utmost importance that assessments of hazards associated with chemicals should be based on test data of assured quality.

Good Laboratory Practice (GLP) is intended to promote the quality and validity of test data. It is a managerial concept covering the organisational process and the conditions under which laboratory studies are planned, performed, monitored, recorded and reported.

The application of GLP is of crucial importance to national authorities entrusted with the responsibility of assessing test data and evaluating chemical hazards. The issue however also has an international dimension. If countries can rely on test data developed in other countries, duplicative testing can be avoided and costs to government and industry saved. Moreover, common principles and procedures for GLP would facilitate the exchange of information and prevent the emergence of non-tariff barriers to trade while enhancing environmental and health protection.

The project on GLP, together with projects on an International Glossary of Key Terms, Confidentiality of Data, and Information Exchange, formed the first phase of the Special Programme on the Control of Chemicals.

The products of the Expert Group are already finding practical application. The Group developed the OECD Principles of GLP (Chapter 2) which have been reviewed in the relevant policy bodies of the Organisation and are now formally recommended for use in Member countries by the OECD Council. Moreover, on 12th May, 1981, Council adopted a Decision on the Mutual Acceptance of Data which states "that data generated in the testing of chemicals in an OECD Member country in accordance with OECD Test Guidelines(*) and OECD Principles of Good Laboratory Practice shall be accepted in other Member countries for purposes of assessment and other uses relating to the protection of man and the environment".

(*) "OECD Guidelines for Testing of Chemicals" (1981 and continuing series) OECD: Paris.

In adopting the Decision, the Council called for work on internationally-harmonized approaches to assuring compliance with the OECD Principles of GLP. In Chapter 3 of this report the national and international elements which bear on this issue are discussed. In Chapter 4, guidelines are proposed for a specific, national aspect of GLP compliance, namely inspections and study audits.

As one of the Expert Group reports prepared under the first phase of the OECD Special Programme on the Control of Chemicals, this document has been derestricted under the authority of the Secretary-General. The views expressed do not necessarily reflect those of OECD or its Member countries; they provide the basis for future considerations of these issues by the OECD Council and its subsidiary bodies.

Jim MacNeill
Director
OECD Environment Directorate

This report was prepared by an Expert Group under the chairmanship of Dr. Carl Morris, United States Environmental Protection Agency. The United States provided secretarial support and the following countries and organisations participated in the work of the Group: Australia, Austria, Belgium, Canada, Denmark, France, Federal Republic of Germany, Greece, Italy, Japan, Netherlands, New Zealand, Norway, Sweden, Switzerland, United Kingdom, United States, the Commission of the European Communities, the World Health Organisation and the International Organisation for Standardisation (ISO/CERTICO).

Also available

CHEMICALS CONTROL LEGISLATION: AN INTERNATIONAL GLOSSARY OF KEY TERMS (1982)
(59 82 02 1) ISBN 92-64-12364-4
£6.80   US$13.50   F68.00

CONFIDENTIALITY OF DATA AND THE CONTROL OF CHEMICALS (1982)
(59 82 03 1) ISBN 92-64-12365-2
£5.00   US$10.00   F50.00

OECD GUIDELINES FOR TESTING OF CHEMICALS (July 1981)
(97 81 05 1) ISBN 92-64-12221-4,   734 pages
£35.00   US$80.00   F360.00

CONTROL OF CHEMICALS IN IMPORTING COUNTRIES (August 1982)
(59 82 01 1) ISBN 92-64-12272-9,   106 pages
£5.40   US$12.00   F54.00

OECD AND CHEMICALS CONTROL (March 1981)
(02 81 03 1) "OECD Background Paper"
£4.00   US$9.00   F40.00

Prices charged at the OECD Publications Office.

THE OECD CATALOGUE OF PUBLICATIONS and supplements will be sent free of charge on request addressed either to OECD Publications Office, 2, rue André-Pascal, 75775 PARIS CEDEX 16, or to the OECD Sales Agent in your country.

TABLE OF CONTENTS

Page

EXECUTIVE SUMMARY 9

Chapter 1: CONSIDERATIONS OF THE OECD EXPERT
GROUP ON GOOD LABORATORY PRACTICE 11

1.1 Introduction 11

1.2 Background to OECD Work on
Good Laboratory Practice 12

1.3 Expert Group on Good
Laboratory Practice 13

1.4 Principles of Good
Laboratory Practice 14

1.5 Implementation of OECD Principles
of Good Laboratory Practice 15

1.6 Guidelines for National GLP
Inspections and Study Audits 17

1.7 References 19

Chapter 2: OECD PRINCIPLES OF GOOD
LABORATORY PRACTICE 23

2.1 Introduction 23

2.2 Definitions of Terms 24

2.3 Good Laboratory Practice
Principles 25

2.4 References 39

Chapter 3: IMPLEMENTATION OF OECD PRINCIPLES
OF GOOD LABORATORY PRACTICE 41

3.1 Introduction 41

3.2 National Approaches to
Implementation 42

3.3 International Recognition
and Co-operation 45

		Page
Chapter 4:	OECD GUIDELINES FOR NATIONAL GLP INSPECTIONS AND STUDY AUDITS	49
4.1	Introduction	49
4.2	Definitions	49
4.3	Procedures	50
Annex I:	MANDATE OF THE GROUP OF EXPERTS [C(78)127(Final) Appendix]	57
Annex II:	EXCERPTS FROM THE DECISION OF THE COUNCIL CONCERNING THE MUTUAL ACCEPTANCE OF DATA IN THE ASSESSMENT OF CHEMICALS [C(81)30 (Final)]	59
Annex III:	EXPERTS WHO PARTICIPATED IN THE WORK OF THE GROUP	61

EXECUTIVE SUMMARY

The Group of experts has identified three essential elements upon which mutual acceptance of data can be based. These elements include:

- the use of the OECD Test Guidelines;
- the application of the OECD Principles of Good Laboratory Practice; and
- the establishment of harmonized national GLP compliance programmes.

In order to attain these objectives, the Expert Group recommends the use of the following chapters in addition to the OECD Test Guidelines.

The background to the Group's work, and the considerations which guided that work, are outlined in Chapter 1.

The OECD Principles of Good Laboratory Practice (Chapter 2) provide guidance to the test facility in order to promote the development of quality data in the testing of chemicals. This document was adopted by the OECD Council on 12th May, 1981, with the recommendation that Member countries apply the OECD Principles of Good Laboratory Practice in the testing of chemicals (see Annex II).

Chapter 3 on implementation, describes an internationally harmonized approach to assuring compliance with the OECD Principles of Good Laboratory Practice. At the national level, it presents essential elements considered necessary for the establishment of an effective national GLP compliance programme. At the international level, it presents a recommendation for international co-operation and recognition in order to foster the concept of the mutual acceptance of data.

The OECD Guidelines for National GLP Inspections and Study Audits (Chapter 4) provide guidance to national inspectors in the performance of inspections and study audits.

CHAPTER 1

CONSIDERATIONS OF THE OECD EXPERT GROUP ON GOOD LABORATORY PRACTICE

1.1: INTRODUCTION

As a result of the awareness that chemicals have the potential to adversely affect human health and the environment, a number of countries have introduced chemicals control legislation during the 1970s requiring the testing of industrial chemicals. This legislation complements existing legislation on the control of specific types of chemicals, such as pharmaceuticals, food additivies, cosmetics, pesticides and explosives. The quality of test data plays a decisive role in the proper functioning of all such legislation and in the exercise of this regulatory power in decision-making concerning the safety of a chemical.

Whenever chemicals are tested, therefore, the test data should be scientifically reliable and of the highest quality. The primary responsibility for producing reliable test data rests with those carrying out and/or commissioning the tests.

Within OECD, a major co-operative effort by Member countries has resulted in the production of OECD Test Guidelines. The Guidelines are state-of-the-art methods for chemicals testing and reflect current scientific knowledge in the relevant fields.

Good laboratory practice is concerned with the quality of test data and the management of laboratory studies. The generation of test data according to currrent scientific knowledge and the best managerial practices has been seen as the best way to promote the international acceptability of test data. OECD has examined both the national and international aspects of good laboratory practice.

This Chapter highlights the issues which were discussed by the OECD Expert Group in the development of this final report, focussing particular attention on national responsibilities and international recognition and co-operation.

1.2: BACKGROUND TO OECD WORK ON GOOD LABORATORY PRACTICE

The earliest regulatory usage of the term "good laboratory practice" (GLP) is found in the New Zealand Testing Laboratory Registration Act of 1982 (1). The Act aimed to establish a coherent national testing policy and set up a Testing Laboratory Registration Council to pursue this goal.

In 1973, the Danish Government adopted similar measures (2).

In 1975, the United States Food and Drug Administration (FDA) found that some studies submitted to it in compliance with the Federal Food, Drug, and Cosmetics Act had not been conducted "in accord with acceptable practices and as such were not of a quality and integrity to assure product safety" (3).

By 1978 final regulations were published which established the United States Food and Drug Administration as the first government agency to require compliance with explicit GLP regulations (4). FDA's action also stimulated consideration of GLP in the United States Environmental Protection Agency (5-7), in other countries (8-11), and in international organisations such as OECD and WHO.

In the same year, representatives of sixteen OECD Member countries and six international organisations met in Stockholm to identify priority items for international attention. Two of the priority items identified at that meeting were (12):

- the development of consistent data requirements and testing methods; and

- the development of consistent standards for good laboratory practice and effective means of enforcing them.

The OECD was designated as the appropriate international organisation for action on these items. The outcome of this international meeting was transformed into the OECD Council Decision Concerning the Special Programme on the Control of Chemicals (13). The Management Committee, was established to supervise work to be carried out under the Special Programme and subsequently set up an Expert Group to work on GLP, with the United States as Lead country. At this time, work was underway on data requirements and testing methods within the OECD Chemicals Testing Programme.

1.3: EXPERT GROUP ON GOOD LABORATORY PRACTICE

The work of the Expert Group falls into two phases which correspond closely to the priority issues identified at Stockholm: guidelines for GLP and means of enforcing them.

Although the Group first met in April 1975, by March 1980 it had completed the first part of its mandate with the production of the OECD Principles of GLP (see Chapter 2 of this book). The Principles provide guidance to test facilities on measures designed to promote the development of quality test data.

On 12th May, 1981, on the proposal of the High Level Meeting of the Chemicals Group (14), endorsed by the Environment Committee, the OECD Council adopted the Decision Concerning the Mutal Acceptance of Data in the Assessment of Chemicals [C(81)30 (Final)] (15). Under the overall objective of internationally harmonizing practices and procedures in chemicals control, the OECD Council decided:

> "that data generated in the testing of chemicals in an OECD country in accordance with OECD Test Guidelines and OECD Principles of Good Laboratory Practive shall be accepted in other Member countries for purposes of assessment and other uses relating to the protection of man and the enviroment."

In support of this Decision, Council recommenced that Member countries apply the OECD Test Guidelines and Principles of GLP when testing chemicals. Further, the Management Committee was instructed to develop internationally harmonized approaches to assure compliance with the OECD Principles of GLP.

In keeping with this instruction, the Expert Group prepared "Implementation of the OECD Principles of Good Laboratory Practice" (see Chapter 3 of this publication). This review presents the elements considered necessary by the Group for the establishment of effective national GLP compliance programmes and their mutual recognition.

Two elements considered necessary by the Group for a national compliance scheme were inspections of laboratory facilities and audits of individual studies. Accordingly, the Group developed OECD Guidelines for National GLP Inspections and Study Audits (see Chapter 4 of this publication). The Guidelines are intended to foster a common understanding and practice in the overall GLP compliance programmes of Member countries and to enhance the mutual recognition of national GLP compliance programmes.

1.4: OECD PRINCIPLES OF GOOD LABORATORY PRACTICE

Following a review of available documentation (4, 8, 16, 17) the Group decided to develop "principles" of good laboratory practice. The principles were intended to serve as guides for laboratories performing tests on chemicals for regulatory purposes. In order to develop such principles, the Group had to decide on their scope.

Thus, the term "Good Laboratory Practice" was defined as follows:

"Good Laboratory Practice (GLP) is concerned with the organisational process and the conditions under which laboratory studies are planned, performed, monitored, recorded and reported."

This definition indicates that GLP is a management tool. It deals with how to organise a test laboratory with the objective of promoting and maintaining the quality of test data. The "OECD Principles of Good Laboratory Practice" are considered to represent the state-of-the-art concerning test laboratory management and practice. Furthermore, Good Laboratory Practice, being a management tool, does not interfere with the exercise of scientific knowledge or practice, but rather complements it.

The Group agreed that the Principles should be applied to the testing of chemicals to obtain data on their properties and on their safety with respect to human health or the environment. Specifically, they should be applicable to the following test areas:

- physico-chemical properties;

- toxicological studies designed to evaluate human health effects (short and long-term);

- ecotoxicological studies designed to evaluate environmental effects (short and long-term);

- ecological studies designed to evaluate environmental chemical fate (transport, biodegradation and bioaccumulation).

The Group also felt strongly that the Principles should be applied to the testing of all chemicals regardless of their use(s) (such as industrial chemicals, pharmaceuticals, cosmetics, food additives, pesticides, etc.) and that they should be applied to laboratory as well as to field studies of chemicals. Although monitoring contains elements of GLP, the Group did not wish to address this type of study at that time.

The Group developed the Principles to meet the requirements for chemicals testing imposed by regulations. In other words, laboratories performing research, which is not intended to be submitted to regulatory authorities, are not required to introduce the OECD Principles of GLP, but for reasons of good management and for the promotion of quality test data they may decide to adhere to these Principles as closely as possible.

The Group considered the relationship between the Principles of GLP and the OECD Test Guidelines. It was intended that the Principles should address only the general aspects of laboratory management and practice, and that any test-specific GLP issues might be addressed in the individual Test Guidelines. However, when some Group members reviewed the OECD Test Guidelines to determine whether additional GLP elements were required, they concluded that the Principles were sufficiently consistent and complete as to cover all GLP elements required by the Test Guidelines. In addition, specific recommendations on the consistency and completeness of various Test Guidelines were submitted to the Review Panel which was at that time finalising the first set of OECD Test Guidelines.

Finally, the Group considered the need to update the Principles of GLP and recommended that practical experience from their application be acquired before addressing this question.

1.5: IMPLEMENTATION OF OECD PRINCIPLES OF GOOD LABORATORY PRACTICE

The Expert Group addressed both the national and international issues involved in the development of internationally-harmonized approaches to assure compliance with the OECD Principles of GLP.

The Group agreed that the implementation of the Principles lies entirely within the responsibility of national governments and will be governed by the legal and/or administrative structure of the respective OECD

Member country. However, the fundamental criteria of a national GLP compliance programme should be mutually agreed and applied consistently within OECD Member countries.

The Group agreed on several criteria for a harmonized national GLP compliance programme. The first of these was the promulgation of the OECD Principles of GLP, and the second, the establishment or designation of a national authority to administer the GLP compliance programme.

The Group agreed that national implementation should aim at a single unified approach to comprise all types of test procedures and all regulated chemicals. However, it was recognised that the establishment of a comprehensive national GLP compliance programme was a major task demanding the commitment of financial and manpower resources. It was therefore expected that countries would proceed step by step, both with respect to the different types of tests and to the different types of chemicals, but with the objective of achieving one unified national approach which avoids multiple assessments under different legal mandates and agencies.

The Group thought that national implementation should include a requirement that each study submitted for regulatory purposes contain a statement by the laboratory that the study was carried out in accordance with the OECD Principles of GLP. It was determined that this statement is necessary to provide information to foreign national authorities concerning the status of the laboratory with respect to GLP compliance. To ensure international recognition of this statement, countries should notify the OECD when they have established their GLP compliance programmes.

The Group further concluded that the mechanism whereby national authorities monitor compliance with the OECD Principles of GLP should be periodic on-site inspections of test facilities and study audits. Due to resource implications, the Group recommended that routine inspections be carried out with a frequency of two to four years, depending on the type of study in question.

The education and experience of national inspectors was emphasized to underscore the importance of their activities. It was thought that governments should be involved in the training and orientation of national inspectors and key governmental personnel. Also, countries implementing the Principles of GLP would determine appropriate education and experience for their laboratory personnel.

Regarding the scope of national GLP compliance schemes, it was agreed that they would extend to all non-clinical testing of chemicals to obtain data which are intended to fulfill regulatory requirements.

At the international level, recommendations for co-operation and mutual recognition were made. The major issues can be summarised as follows:

- the establishment of national GLP compliance programmes is seen as an activity supportive to and covered by the Decision of the OECD Council Concerning the Mutual Acceptance of Data in the Assessment of Chemicals [C(81)30 (Final)]; and

- notification to the OECD, when Member countries have established a national GLP compliance programme, is seen as a formal step.

Furthermore, to foster greater confidence among Member countries concerning the quality of test data, a multilateral mechanism for co-operation was recommended. This multilateral mechanism should comprise at least four basic elements:

(i) full and direct communication between national authorities regarding administrative and GLP compliance matters;

(ii) the right to request information from another country in exceptional situations where the recipient country has reasonable doubts about the quality of the data submitted to it;

(iii) the right of a country to participate in an inspection and/or study audit as an observer; and

(iv) the establishment of an international GLP forum which is intended to provide an opportunity for officials of the national competent authorities to discuss technical and administrative matters.

1.6: GUIDELINES FOR NATIONAL GLP INSPECTIONS AND STUDY AUDITS

The Group agreed that one of the elements of a harmonized national GLP complaince programme was the elaboration of procedures for laboratory inspections and study audits to provide guidance for national inspectors. The Guidelines for National GLP Inspections and Study Audits developed by the Group are intended to provide common guidance to national inspectors, thus fostering mutual understanding and common practice in the overall GLP compliance programme of Member countries and enhancing the mutual recognition of national GLP compliance programmes.

Laboratory inspections and study audits are two integral parts of an inspection programme and therefore are dealt with together. In developing Chapter 4, the Group agreed that laboratory inspection and study audit have the following meaning:

Inspection means an on-site examination of the test facility and its procedures in order to determine that these are in compliance with the OECD Principles of GLP.

Study audit means a comparison of raw data and associated records with the final or interim report of an individual study in order to determine that the study was carried out in accordance with the study plan and standard operating procedures.

The inspection guidelines were structured by the Group in the same order as the OECD Principles of GLP to emphasize the step-by-step procedure which should be followed by the inspector. The inspection report should, in turn, be structured in the same way.

1.7: REFERENCES

(1) New Zealand   Testing Laboratory
                       Registration Act 1972
                       No. 36, 20th October 1972.

(2) Denmark   Danish National Testing
                       Board Act No. 144
                       21st March, 1973.

(3) United States   Food and Drug Administration
                       Non-Clinical Laboratory
                       Studies
                       Proposed Regulations for
                       Good Laboratory Practice
                       19th November, 1976,
                       FR 41(225), pp. 51206-51230.

(4) United States   Food and Drug Administration
                       Non-Clinical Laboratory
                       Studies
                       Good Laboratory Practice
                       Regulations
                       Final Rule
                       22nd December, 1978
                       FR 43(247), pp. 59986-60020.

(5) United States   Environmental Protection
                       Agency
                       Toxic Substances Control Act
                       Good Laboratory Practice
                       Standards for Health Effects
                       Proposed Rule
                       9th May, 1979
                       FR 44(91), pp. 27362-27375.

(6) United States   Environmental Protection
                       Agency
                       Toxic Substances Control Act
                       Proposed Good Laboratory
                       Practice
                       Standards for Physical,
                       Chemical, Persistence, and
                       Ecological Effects Testing
                       21st November, 1980
                       FR 45 (227), pp. 77353-77365.

(7)	United States	Environmental Protection Agency Pesticide Programs Guidelines for Registering Pesticides in the United States Proposed Good Laboratory Practice Guidelines for Toxicology Testing 18th April, 1980 FR 45 (77), pp. 26373-26385.
(8)	The Netherlands	Proposed Regulation for Good Laboratory Practice in the Netherlands Non-clinical Toxicity Studies 1979 (unpublished).
(9)	Switzerland	Interkantonale Kontrollstelle für Heilmittel (IKS), Bern Wegleitung der IKS betreffend Gute Laboratoriumspraxis für nichtklinische Laborversuche Bern, 28th April, 1980.
(10)	United Kingdom	Notification of New Substances Draft Regulation and Approved Code of Practice Consultative Document Health and Safety Commission London 1981.
(11)	Japan	Good Laboratory Practice for Pharmaceuticals (Proposed) Ministry of Health and Welfare Tokyo, 22nd July, 1981.
(12)	Stockholm	Proceedings of the April 1978 International Meeting on the Control of Toxic Substances with Special Regard to Environmental Chemicals ed. by the Swedish National Products Control Board Stockholm, April 1978 Conclusions.

(13)   OECD   Decision of the Council
                           Concerning the Special
                           Programme on the Control of
                           Chemicals [C(78)127(Final)]
                           Paris, 16th October, 1978.

(14)   OECD   OECD and Chemicals Control
                           The High Level Meeting of the
                           Chemicals Group of the
                           Environment Committee, 1980
                           Paris (1981)

(15)   OECD   Decision of the OECD Council
                           Concerning the Mutual
                           Acceptance of Data in the
                           Assessment of Chemicals
                           [C(81)30(Final)].

(16)   ECETOC   European Chemical Industry
                           Ecology and Toxicology
                           Centre Good Laboratory
                           Practice Monograph No.1
                           Brussels, October 1979.

(17)   ISO   ISO Guide 25 1978-10-01
                           Guidelines for Assessing the
                           Technical Competence of
                           Testing Laboratories
                           International Organisation
                           for Standardisation, Geneva
                           1978.

CHAPTER 2

OECD PRINCIPLES OF GOOD LABORATORY PRACTICE

2.1: INTRODUCTION

Preface

A number of OECD Member countries have recently passed legislation to control chemical substances and others are about to do so. This legislation usually requires the manufacturer to perform laboratory studies and to submit the results of these studies to a governmental authority for assessment of the potential hazard to human health and the environment.

Government and industry are increasingly concerned with the quality of studies upon which hazard assessments are based. As a consequence, several OECD Member countries have established, or plan to establish, criteria for the performance of these studies.

To avoid different schemes of implementation that could impede international trade in chemicals, OECD Member countries have recognised the unique opportunity for international harmonization of test methods and good laboratory practices.

During 1979-80, an international group of experts established under the Special Programme on the Control of Chemicals developed this document concerning the Principles of Good Laboratory Practice (GLP) utilising common managerial and scientific practices and experience from various national and international sources.

The purpose of these Principles of Good Laboratory Practice is to promote the development of quality test data. Comparable quality of test data forms the basis for the mutual acceptance of test data among countries.

If individual countries can confidently rely on test data developed in other countries, duplicative testing can be avoided, thereby economising test costs and time. The application of these Principles should help avoid the creation of technical barriers to trade and further improve the protection of human health and the environment.

Scope

These Principles of Good Laboratory Practice should be applied to the testing of chemicals to obtain data on their properties and/or their safety with respect to human health or the environment. These data may be developed for the purpose of meeting regulatory requirements. Studies covered by Good Laboratory Practice also include work conducted in field studies.

2.2: DEFINITIONS OF TERMS

Good Laboratory Practice

(1) Good Laboratory Practice (GLP) is concerned with the organisational process and the conditions under which laboratory studies are planned, performed, monitored, recorded, and reported.

Terms concerning the organisation of a test facility

(1) Test facility means the persons, premises, and operational unit(s) that are necessary for conducting the study.

(2) Study Director means the individual responsible for the overall conduct of the study.

(3) Quality Assurance Programme means an internal control system designed to ascertain that the study is in compliance with these Principles of Good Laboratory Practice.

(4) Standard Operating Procedures (SOPs) mean written procedures which describe how to perform certain routine laboratory tests or activities normally not specified in detail in study plans or test guidelines.

(5) Sponsor means a person(s) or entity who commissions and/or supports a study.

Terms concerning the study

(1) Study means an experiment or set of experiments in which a test substance is examined to obtain data on its properties and/or its safety with respect to human health and the environment.

(2) Study plan means a document which defines the entire scope of the study.

(3) **OECD Test Guideline** means a test guideline which the OECD has recommended for use in its Member countries.

(4) **Test system** means any animal, plant, microbial, as well as other cellular, sub-cellular, chemical, or physical system or a combination thereof used in a study.

(5) **Raw data** means all original laboratory records and documentation, or verified copies thereof, which are the result of the original observations and activities in a study.

(6) **Specimen** means any material derived from a test system for examination, analysis, or storage.

Terms concerning the test substance

(1) **Test substance** means a chemical substance or a mixture which is under investigation.

(2) **Reference substance (control substance)** means any well-defined chemical substance or any mixture other than the test substance used to provide a basis for comparison with the test substance.

(3) **Batch** means a specific quantity or lot of a test or reference substance produced during a defined cycle of manufacture in such a way that it could be expected to be of a uniform character and should be designated as such.

(4) **Vehicle (carrier)** means any agent which serves as a carrier used to mix, disperse, or solubilise the test or reference substance to facilitate the administration to the test system.

(5) **Sample** means any quantity of the test or reference substance.

2.3: GOOD LABORATORY PRACTICE PRINCIPLES

TEST FACILITY ORGANISATION AND PERSONNEL

Management's Responsibilities

(1) Test facility management should ensure that the Principles of Good Laboratory Practice are complied with in the test facility.

(2) At a minimum it should:

    (a) ensure that qualified personnel, appropriate facilities, equipment, and materials are available;

    (b) maintain a record of the qualifications, training, experience and job description for each professional and technical individual;

    (c) ensure that personnel clearly understand the functions they are to perform and, where necessary, provide training for these functions;

    (d) ensure that health and safety precautions are applied according to national and/or international regulations;

    (e) ensure that appropriate Standard Operating Procedures are established and followed;

    (f) ensure that there is a Quality Assurance Programme with designated personnel;

    (g) where appropriate, agree to the study plan in conjunction with the sponsor;

    (h) ensure that amendments to the study plan are agreed upon and documented;

    (i) maintain copies of all study plans;

    (j) maintain a historical file of all Standard Operating Procedures;

    (k) for each study ensure that a sufficient number of personnel is available for its timely and proper conduct;

    (l) for each study designate an individual with the appropriate qualifications, training, and experience as the Study Director before the study is initiated. If it is necessary to replace a Study Director during a study, this should be documented;

    (m) ensure that an individual is identified as responsible for the management of the archives.

Study Director's Responsibilities

(1) The Study Director has the responsibility for the overall conduct of the study and for its report.

(2) These responsibilities should include, but not be limited to, the following functions:

    (a) should agree to the study plan;

    (b) ensure that the procedures specified in the study plan are followed, and that authorisation for any modification is obtained and documented together with the reasons for them;

    (c) ensure that all data generated are fully documented and recorded;

    (d) sign and date the final report to indicate acceptance of responsibility for the validity of the data and to confirm compliance with these Principles of Good Laboratory Practice;

    (e) ensure that after termination of the study, the study plan, the final report, raw data and supporting material are transferred to the archives.

Personnel Responsibilities

(1) Personnel should exercise safe working practice. Chemicals should be handled with suitable caution until their hazard(s) has been established.

(2) Personnel should exercise health precautions to minimise risk to themselves and to ensure the integrity of the study.

(3) Personnel known to have a health or medical condition that is likely to have an adverse effect on the study should be excluded from operations that may affect the study.

QUALITY ASSURANCE PROGRAMME

General

(1) The test facility should have a documented quality assurance programme to ensure that studies performed are in compliance with these Principles of Good Laboratory Practice.

(2) The quality assurance programme should be carried out by an individual or by individuals designated by and directly responsible to management and who are familiar with the test procedures.

(3) This individual(s) should not be involved in the conduct of study being assured.

(4) This individual(s) should report any findings in writing directly to management and to the Study Director.

Responsibilities of the Quality Assurance Personnel

(1) The responsibilities of the quality assurance personnel should include, but not be limited to, the following functions:

    (a) ascertain that the study plan and Standard Operating Procedures are available to personnel conducting the study;

    (b) ensure that the study plan and Standard Operating Procedures are followed by periodic inspections of the test facility and/or by auditing the study in progress. Records of such procedures should be retained.

    (c) promptly report to management and the Study Director unauthorised deviations from the study plan and from Standard Operating Procedures;

    (d) review the final reports to confirm that the methods, procedures, and observations are accurately described, and that the reported results accurately reflect the raw data of the study;

    (e) prepare and sign a statement, to be included with the final report, which specifies the dates inspections were made and the dates any findings were reported to management and to the Study Director.

FACILITIES

General

(1)   The test facility should be of suitable size, construction and location to meet the requirements of the study and minimise disturbances that would interfere with the validity of the study.

(2)   The design of the test facility should provide an adequate degree of separation of the different activities to assure the proper conduct of each study.

Test System Facilities

(1)   The test facility should have a sufficient number of rooms or areas to assure the isolation of test systems and the isolation of individual projects, involving substances known or suspected of being biohazardous.

(2)   Suitable facilities should be available for the diagnosis, treatment and control of diseases, in order to ensure that there is no unacceptable degree of deterioration of test systems.

(3)   There should be storage areas as needed for supplies and equipment. Storage areas should be separated from areas housing the test systems and should be adequately protected against infestation and contamination. Refrigeration should be provided for perishable commodities.

Facilities for Handling Test and Reference Substances

(1)   To prevent contamination or mix-ups, there should be separate areas for receipt and storage of the test and reference substances, and mixing of the test substances with a vehicle.

(2)   Storage areas for the test substances should be separate from areas housing the test systems and should be adequate to preserve identity, concentration, purity, and stability, and ensure safe storage for hazardous substances.

Archive Facilities

(1)   Space should be provided for archives for the storage and retrieval of raw data, reports, samples, and specimens.

Waste Disposal

(1) Handling and disposal of wastes should be carried out in such a way as not to jeopardise the integrity of studies in progress.

(2) The handling and disposal of wastes generated during the performance of a study should be carried out in a manner which is consistent with pertinent regulatory requirements. This would include provision for appropriate collection, storage, and disposal facilities, decontamination and transportation procedures, and the maintenance of records related to the preceding activities.

APPARATUS, MATERIAL, AND REAGENTS

Apparatus

(1) Apparatus used for the generation of data, and for controlling environmental factors relevant to the study should be suitably located and of appropriate design and adequate capacity.

(2) Apparatus used in a study should be periodically inspected, cleaned, maintained, and calibrated according to Standard Operating Procedures. Records of procedures should be maintained.

Material

(1) Apparatus and materials used in studies should not interfere with the test systems.

Reagents

(1) Reagents should be labelled, as appropriate, to indicate source, identity, concentration, and stability information and should include the preparation date, earliest expiration date, specific storage instructions.

TEST SYSTEMS

Physical/Chemical

(1) Apparatus used for the generation of physical/chemical data should be suitably located and of appropriate design and adequate capacity.

(2) Reference substances should be used to assist in ensuring the integrity of the physical/chemical test systems.

Biological

(1) Proper conditions should be established and maintained for the housing, handling and care of animals, plants, microbial as well as other cellular and sub-cellular systems, in order to ensure the quality of the data.

(2) In addition, conditions should comply with appropriate national regulatory requirements for the import, collection, care and use of animals, plants, microbial as well as other cellular and sub-cellular systems.

(3) Newly received animal and plant test systems should be isolated until their health status has been evaluated. If any unusual mortality or morbidity occurs, this lot should not be used in studies and, when appropriate, humanely destroyed.

(4) Records of source, date of arrival, and arrival condition should be maintained.

(5) Animal, plant, microbial, and cellular test systems should be acclimatised to the test environment for an adequate period before a study is initiated.

(6) All information needed to properly identify the test systems should appear on their housing or containers.

(7) The diagnosis and treatment of any disease before or during a study should be recorded.

TEST AND REFERENCE SUBSTANCES

Receipt, Handling, Sampling and Storage

(1) Records including substance characterisation, date of receipt, quantities received, and used in studies should be maintained.

(2) Handling, sampling, and storage procedures should be identified in order that the homogeneity and stability is assured to the degree possible and contamination or mixup are precluded.

(3) Storage container(s) should carry identification information, earliest expiration date, and specific storage instructions.

Characterisation

(1) Each test and reference substance should be appropriately identified [e.g. code, chemical abstract number (CAS), name].

(2) For each study, the identity, including batch number, purity, composition, concentrations, or other characterisations to appropriately define each batch of the test or reference substances should be known.

(3) The stability of test and reference substances under conditions of storage should be known for all studies.

(4) The stability of test and reference substances under the test conditions should be known for all studies.

(5) If the test substance is administered in a vehicle, Standard Operating Procedures should be established for testing the homogeneity and stability of the test substance in that vehicle.

(6) A sample for analytical purposes from each batch of test substance should be retained for studies in which the test substance is tested longer than four weeks.

STANDARD OPERATING PROCEDURES

General

(1) A test facility should have written Standard Operating Procedures approved by management that are intended to ensure the quality and integrity of the data generated in the course of the study.

(2) Each separate laboratory unit should have immediately available Standard Operating Procedures relevant to the activities being performed therein. Published text books, articles and manuals may be used as supplements to these Standard Operating Procedures.

Application

(1) Standard Operating Procedures should be available for, but not be limited to, the following categories of laboratory activities. The details given under each heading are to be considered as illustrative examples.

(a) <u>Test and reference substance</u>
Receipt, identification, labelling, handling, sampling, and storage.

(b) <u>Apparatus and reagents</u>
Use, maintenance, cleaning, calibration of measuring apparatus and environmental control equipment; preparation of reagents.

(c) <u>Record keeping, reporting, storage, and retrieval</u>
Coding of studies, data collection, preparation of reports, indexing systems, handling of data, including the use of computerised data systems.

(d) <u>Test system (where appropriate)</u>

    (i) Room preparation and environmental room conditions for the test system.

    (ii) Procedures for receipt, transfer, proper placement, characterisation, identification and care of test system.

    (iii) Test system preparation, observation examinations, before, during and at termination of the study.

    (iv) Handling of test system individuals found moribund or dead during the study.

    (v) Collection, identification and handling of specimens including necropsy and histopathology.

(e) <u>Quality assurance procedures</u>
Operation of quality assurance personnel in performing and reporting study audits, inspections, and final study report reviews.

(f) <u>Health and safety precautions</u>
As required by national and/or international legislation or guidelines.

PERFORMANCE OF THE STUDY

Study Plan

(1) For each study, a plan should exist in a written form prior to initiation of the study.

(2) The study plan should be retained as raw data.

(3) All changes, modifications, or revisions of the study plan, as agreed to by the Study Director, including justification(s), should be documented, signed and dated by the Study Director, and maintained with the study plan.

Content of the Study Plan

The study plan should contain, but not be limited to the following information:

(1) Identification of the study, the test and reference substance

    (a) A descriptive title;

    (b) A statement which reveals the nature and purpose of the study;

    (c) Identification of the test substance by code or name (IUPAC; CAS number, etc);

    (d) The reference substance to be used.

(2) Information concerning the sponsor and the test facility

    (a) Name and address of the Sponsor;

    (b) Name and address of the Test Facility;

    (c) Name and address of the Study Director;

(3) Dates

    (a) The date of agreement to the study plan by signature of the Study Director, and when appropriate, of the sponsor and/or the test facility management;

    (b) The proposed starting and completion dates.

(4) Test methods

    (a) Reference to OECD Test Guideline or other test guideline to be used.

(5) Issues (where applicable)

    (a) The justification for selection of the test system;

- (b) Characterisation of the test system, such as the species, strain, substrain, source of supply, number, body weight range, sex, age, and other pertinent information;

- (c) The method of administration and the reason for its choice;

- (d) The dose levels and/or concentration(s), frequency, duration of administration;

- (e) Detailed information on the experimental design, including a description of the chronological procedure of the study, all methods, materials and conditions, type and frequency of analysis, measurements, observations and examinations to be performed.

(6) <u>Records</u>

- (a) A list of records to be retained.

Conduct of the Study

(1) A unique identification should be given to each study. All items concerning this study should carry this identification.

(2) The study should be conducted in accordance with the study plan.

(3) All data generated during the conduct of the study should be recorded directly, promptly, accurately, and legibly by the individual entering the data. These entries should be signed or initialled and dated.

(4) Any change in the raw data should be made so as not to obscure the previous entry, and should indicate the reason, if necessary, for change and should be identified by date and signed by the individual making the change.

(5) Data generated as a direct computer input should be identified at the time of data input by the individual(s) responsible for direct data entries. Corrections should be entered separately by the reason for change, with the date and the identity of the individual making the change.

REPORTING OF STUDY RESULTS

General

(1) A final report should be prepared for the study.

(2) The use of Standard International Units is recommended.

(3) The final report should be signed and dated by the Study Director.

(4) If reports of principal scientists from co-operating disciplines are included in the final report, they should sign and date them.

(5) Corrections and additions to a final report should be in the form of an amendment. The amendment should clearly specify the reason for the corrections or additions and should be signed and dated by the Study Director and by the principal scientist from each discipline involved.

Content of the Final Report

The final report should include, but not be limited to, the following information:

(1) Identification of the study, the test and reference substance

(a) A descriptive title;

(b) Identification of the test substance by code or name (IUPAC; CAS number, etc);

(c) Identification of the reference substance by chemical name;

(d) Characterisation of the test substance including purity, stability and homogeneity.

(2) Information concerning the test facility

(a) Name and address;

(b) Name of the Study Director;

(c) Name of other principal personnel having contributed reports to the final report.

(3) <u>Dates</u>

    (a) Dates on which the study was initiated and completed.

(4) <u>Statement</u>

    (a) A Quality Assurance statement certifying the dates inspections were made and the dates any findings were reported to management and to the Study Director;

(5) <u>Description of materials and test methods</u>

    (a) Description of methods and materials used;

    (b) Reference to OECD Test Guidelines or other test guidelines.

(6) <u>Results</u>

    (a) A summary of results;

    (b) All information and data required in the study plan;

    (c) A presentation of the results, including calculations and statistical methods;

    (d) An evaluation and discussion of the results and, where appropriate, conclusions.

(7) <u>Storage</u>

    (a) The location where all samples, specimens, raw data, and the final report are to be stored.

<u>STORAGE AND RETENTION OF RECORDS AND MATERIAL</u>

<u>Storage and Retrieval</u>

(1) Archives should be designed and equipped for the accommodation and the secure storage of:

    (a) The study plans;

    (b) The raw data;

    (c) The final reports;

    (d) The reports of laboratory inspections and study audits performed according to the Quality Assurance Programme;

(e) Samples and specimens.

(2) Material retained in the archives should be indexed so as to facilitate orderly storage and rapid retrieval.

(3) Only personnel authorised by management should have access to the archives. Movement of material in and out of the archives should be properly recorded.

Retention

(1) The following should be retained for the period specified by the appropriate authorities:

(a) The study plan, raw data, samples, specimens, and the final report of each study;

(b) Records of all inspections and audits performed by the Quality Assurance Programme;

(c) Summary of qualifications, training, experience and job descriptions of personnel;

(d) Records and reports of the maintenance and calibration of equipment;

(e) The historical file of Standard Operating Procedures.

(2) Samples and specimens should be retained only as long as the quality of the preparation permits evaluation.

(3) If a test facility or an archive contracting facility goes out of business and has no legal successor, the archive should be transferred to the archives of the sponsor(s) of the study(s).

2.4: REFERENCES

The following references are provided as further guidance to the OECD Principles of Good Laboratory Practice.

(1) U.S. Food and Drug Administration,
Non-Clinical Laboratory Studies, Good Laboratory Practice Regulations,
U.S. Federal Register. Vol. 41, No. 225, 19 November 1976, pp. 51206-51226, (Proposed Regulations) and Vol. 43, No. 247, 22 December 1978, pp. 59986-60020, (Final Rule).

(2) U.S. Environmental Protection Agency,
Good Laboratory Practice Standards for Health Effects, U.S. Federal Register Vol. 44, No. 91, 9 May 1979, pp. 27334-27375, (Proposed Rule).

(3) Environmental Health Criteria 6,
Principles and Methods for Evaluating Toxicity of Chemicals, Part I
World Health Organisation, Geneva, 1978.

(4) Good Laboratory Practice, European Chemical Industry Ecology and Toxicology Centre (ECETOC), Monograph No. 1,
Brussels, October 1979.

(5) Good Laboratory Practice,
ed. by G.E. Paget,
MTP Press Limited,
Lancaster 1979.

(6) Stanley L. Inhorn, Quality Assurance Practices for Health Laboratories. American Public Health Association, Washington 1978.

(7) Quality Control in Toxicology,
ed. by G.E. Paget,
MTP Press Limited,
Lancaster 1977.

(8) Standard Operating Procedures in Toxicology,
ed. by G.E. Paget and R. Thomson,
MTP Press Limited,
Lancaster 1979.

(9) Standard Operating Procedures in Pathology,
ed. by G.E. Paget and R. Thomson,
MTP Press Limited,
Lancaster 1979.

(10)   Proceedings of the Colloquium
        Quality Assurance of Toxicological Data
        Luxembourg 11-13 December 1979
        ed. Commission of the European Communities and
        U.S. Environmental Protection Agency, forthcoming
        1980

(11)   D.A. Lowe, A Guide to International
        Recommendations on Names and Symbols for Quantities and on Units of Measurement.
        World Health Organisation, Geneva 1975.

(12)   Guide for the Care and Use of Laboratory
        Animals, Institute of Laboratory Animal Resources, National Research Council, U.S. Dept.
        Health, Education and Welfare (DHEW Publication
        No. (NIH) 78-23 Revised 1978).

(13)   Long-Term Holding of Laboratory Rodents. Prepared by a Committee of the Institute of
        Laboratory Animal Resources, Assembly of Life
        Sciences, National Research Council (Institute
        of Laboratory Animal Resource News, Vol. XIX,
        No. 4, 1976).

(14)   Le Manuel du Technicien Animalier, préparé par
        le personnel de la Division des ressources
        animales, Direction général de la protection de
        la Santé, Ministère de la Santé National et du
        Bien-être, Canada, Ottawa 1975.

(15)   Royal Australian Chemical Institute (RACI),
        Ad Hoc Hazardous Research Materials Committee,
        Draft Code for the Disposal of Laboratory Wastes,
        Chemistry in Australia, Vol. 46, No. 8,
        pp. 344-347, August 1979.

(16)   Soins a prodiger aux animaux d'expérience,
        Guide pour le Canada, Conseil Canadien de
        Protection des Animaux;
        Guide to the care and uses of experimental
        animals, Vol 1,
        Canadian Council on Animal Care,
        151 Slater, Ottawa K1P 5H3, Canada.

CHAPTER 3

IMPLEMENTATION OF THE OECD PRINCIPLES OF GOOD LABORATORY PRACTICE

3.1: INTRODUCTION

Objective

In accordance with the "Decision of the OECD Council Concerning the Mutual Acceptance of Data in the Assessment of Chemicals" [C(81)30 (Final)](*), the purposes of this document are:

- to promote the implementation of comparable national programmes among OECD Member countries for compliance with the OECD Principles of Good Laboratory Practice (GLP), in order to foster mutual acceptance of test data with respect to the potential hazards of chemicals to human health and the environment; and

- to suggest ways for OECD Member countries to co-operate internationally in order to achieve harmonized approaches to assure compliance with the OECD Principles of Good Laboratory Practice.

Scope of Implementation

The OECD Principles of Good Laboratory Practice can be applied to all non-clinical testing of chemicals to obtain data which are intended to fulfill regulatory requirements. To encourage the establishment of a unified national GLP compliance programme within each Member country, it is recommended that the OECD Principles of Good Laboratory Practice and the OECD Guidelines for National GLP Laboratory Inspections and Study Audits be implemented for the testing of all regulated chemicals including industrial chemicals, pharmaceuticals, and pesticides.

(*) Relevant excerpts from the Council Decision are attached as Annex II.

The Concept

Procedures for controlling conformity to standards have always been of concern to government and industry, especially when they have an impact on international trade. Thus, national programmes which are established by OECD Member countries to ensure compliance with OECD Principles of Good Laboratory Practice should include the elements described below in order to promote the quality of test data and the comparability of GLP compliance programmes.

Health and environmental test data generated in one country may be submitted to another country to satisfy regulatory reporting requirements for chemicals. If individual countries can confidently rely on test data developed in other countries, duplicative testing can be avoided, thereby introducing significant savings in costs, time, and number of test animals. Harmonization in these areas will foster international acceptance of data and avoid duplication of testing, thereby avoiding non-tariff barriers to trade.

The implementation of the OECD Principles of Good Laboratory Practice lies entirely within the responsibility of national governments and will be governed by the legal and/or administrative structure of the respective OECD Member country.

It is expected that each Member country will develop its own priorities for implementation, both for the testing of different types of chemicals and for different types of testing, but with the objective of achieving a unified national approach to the implementation of the OECD Principles of Good Laboratory Practice. To be internationally harmonized, national GLP compliance programmes should include the fundamental elements described in Section 3.2, below.

3.2: NATIONAL APPROACHES TO IMPLEMENTATION OF OECD PRINCIPLES OF GOOD LABORATORY PRACTICE

Adoption of OECD Principles of Good Laboratory Practice

OECD Principles of Good Laboratory Practice can be incorporated into laws, regulations, codes of practice or recommendation depending upon the national, legal and/or administrative practices existing in each OECD Member country.

National Responsibilities

National authorities have the responsibility to implement a GLP compliance programme which includes provisions for taking action in cases of non-compliance. These compliance programmes will vary according to administrative requirements of various countries, however, they must provide that national authorities have final responsibility in all programme and compliance matters. The administration of national GLP compliance programmes should be described in written documents.

A national GLP compliance programme must include a requirement that each study submitted for regulatory purposes contain a statement by the laboratory that the study was carried out in accordance with the OECD Principles of Good Laboratory Practice or with national regulations or equivalents conforming to these Principles.

National Compliance Programmes

The principal mechanisms whereby national authorities monitor compliance with the OECD Principles of Good Laboratory Practice are inspections of laboratories and study audits. The "OECD Guidelines for National GLP Inspections and Study Audits" should be used by inspectors when conducting laboratory inspections and/or study audits. The frequency of inspections and the frequency of study audits are independent decisions to be made by the national authorities in formulating their GLP compliance programmes.

An inspection is an on-site examination of the test facility and its procedures to determine that these are in compliance with the OECD Principles of Good Laboratory Practice. An inspection results in a written report which identifies areas of compliance and any deficiencies.

Study audits are based on the premise that evidence of satisfactory performance can be found in the records made during the conduct of the study. A study audit is a comparison of raw data and associated records with the report of an individual study. The purpose of the audit is to verify the accurate recording of the data in the report and to determine whether the study was carried out in accordance with the study plan and standard operating procedures. A study audit results in a written report.

Frequency of Inspections

The dynamic characteristics of laboratories, such as changes in personnel, modification of facilities, laboratory procedures or management structure, make it necessary to maintain a current assessment of such changes and the way they may have an impact on the quality of the test data. In order to monitor the current status of these factors, it is essential that laboratory inspections be carried out at regular intervals. The frequency of inspections will depend upon such factors as the resources and compliance priorities of the national authorities as well as the types of testing conducted by the laboratory. In general, an inspection frequency of two to four years is recommended.

Personnel and Training

National authorities must use properly trained personnel who are competent to assess the compliance of laboratories with the OECD Principles of Good Laboratory Practice as well as to administer the GLP compliance programme. Such personnel should have suitable education and/or experience and be thoroughly acquainted with the national GLP compliance programme and its administration.

National authorities must use properly trained personnel for inspecting test facilities and auditing studies. They should also be familiar with related scientific and technical issues.

The training of inspectors is a responsibility of national authorities and should include periodic updating. Training could be carried out by drawing on the experience available in government, industry, and academia.

National Consequences of Compliance and Non-Compliance

When an inspection of a test facility and an audit of studies by a national authority indicate a satisfactory level of compliance with the OECD Principles of Good Laboratory Practice, then that test facility and the data from studies performed by it should be acceptable to the national authority. The consequences of non-compliance should be determined by each national authority. Within any GLP compliance programme, there should be provisions for actions which may be taken by national authorities for failure to comply with the OECD Principles of Good Laboratory Practice and to achieve correction of deficiencies.

It is desirable that a mechanism exist for consultation and/or appeal by a test facility regarding administrative decisions on compliance with the OECD Principles of Good Laboratory Practice.

3.3: INTERNATIONAL RECOGNITION AND CO-OPERATION

Mutual Acceptance of Test Data

The Decision of the OECD Council Concerning the Mutual Acceptance of Data in the Assessment of Chemicals [C(81)30 (Final)] calls for harmonized approaches to assuring compliance with the OECD Principles of Good Laboratory Practice. Therefore, national GLP compliance programmes should contain the fundamental elements described above.

The establishment of effective national GLP compliance programmes along with the application of the OECD Principles of Good Laboratory Practice and the use of the OECD Guidelines for Testing of Chemicals provide the assurance necessary for the mutual acceptance of test data. However, there is also a need for an international mechanism as outlined in the following sections for recognising the comparability of GLP compliance programmes of each OECD Member country.

Mutual Recognition of National GLP Compliance Programmes

Prior to the acceptance of the OECD Principles of Good Laboratory Practice, agreements on Good Laboratory Practice were developed on the basis of bilateral memoranda of understanding between national competent authorities of several countries. These efforts have provided useful guidance and have identified, in some measure, the benefits which now may flow from a multilateral mechanism.

A multilateral mechanism for recognising and fostering the development of comparable national GLP compliance programmes will obviate the need for bilateral agreements and promote the objective of harmonization in the field of Good Laboratory Practice. Such a mechanism can be established as other OECD countries institute harmonized national GLP compliance programmes.

Implementation of the OECD Council Decision Concerning the Mutual Acceptance of Data in the Assessment of Chemicals

In order to implement the OECD Principles of Good Laboratory Practice within the Decision of the OECD Council Concerning the Mutual Acceptance of Data in the Assessment of Chemicals Member countries should notify the OECD that they have established a Good Laboratory Practice compliance programme consistent with the provisions in section 3.2 above. In particular, this notification should include the following information:

- the identification of the national authority responsible for the administration of the national GLP compliance programme and the designation of a contact point for international communications;

- a description of the national GLP compliance programme with measures taken to implement the GLP compliance programme, administrative procedures, training and orientation for inspectorate personnel, and audit/inspection procedures.

Upon such notification, Member countries can confidently accept for purposes of assessment, data generated in accordance with the OECD Principles of Good Laboratory Practice. It is recommended that the OECD periodically inform Member countries of all such notifications.

Consultation and Verification

There should be full and direct communication among national authorities regarding administrative and compliance matters of their respective national GLP compliance programmes. Regular communication among national authorities will encourage harmonized interpretation and application of the OECD Principles of Good Laboratory Practice and foster greater confidence among Member countries concerning the quality of test data generated in another country.

Although the OECD Council Decision on the Mutual Acceptance of Data in the Assessment of Chemicals provides for the acceptance of data generated in accordance with the OECD Test Guidelines and the OECD Principles of Good Laboratory Practice, there may be exceptional situations in which the recipient country has reasonable doubts about the quality of the data submitted. Therefore, provision needs to be made for the recipient country to identify and to justify its concerns and to request information from the country of origin about the test data and results of inspections or study audits. This request for information may result in a re-inspection or study audit by the national authority from the country of origin. In accordance with national laws and regulations, the national authority of the country of origin can invite, with the consent of the sponsor and the test facility, an official representative of the requesting country to participate as an observer in an inspection or study audit.

The reports of the inspections or study audits relevant to the request will be provided to the requesting country. Due consideration should be given to the confidentiality issues arising from the transmission of these reports.

International GLP Forum

It is recommended that the OECD should co-ordinate and facilitate the establishment of an international GLP forum, open to all Member countries, in which officials of the national competent authorities could meet at least once a year to

- discuss technical and administrative matters arising from the implementation of the OECD Principles of Good Laboratory Practice and the OECD Guidelines for National GLP Inspections and Study Audits;

- promote co-operation between competent national authorities;

- exchange information on the training of inspectors; and

- promote for inspectors, periodic seminars dealing with various aspects of the OECD Principles of Good Laboratory Practice and the OECD Guidelines for National GLP Inspections and Study Audits.

CHAPTER 4

OECD GUIDELINES FOR NATIONAL GLP INSPECTIONS AND STUDY AUDITS

4.1: INTRODUCTION

The inspection programme is intended to ascertain proper study conduct and assure that the resulting data are of adequate quality to permit hazard assessment. This document provides guidelines for inspections and study audits that are mutually acceptable to the OECD Member countries. Such an inspection programme encompasses laboratory inspections and study audits conducted as a part of an overall GLP compliance programme. The inspection programme results in reports which describe the degree of adherence of test facilities to the OECD Principles of Good Laboratory Practice.

4.2: DEFINITIONS

The definition of terms in the "OECD Principles of Good Laboratory Practice" are applicable to this document. In addition, the following definitions apply to this document.

>Inspection means an on-site examination of the test facility and its procedures in order to determine that these are in compliance with the OECD Principles of Good Laboratory Practice.

>Inspection Report means the report which is written after the inspection and which identifies areas of compliance and any deficiencies of the test facility and/or the study audited.

>Inspector means the person who performs the laboratory inspection and/or the study audit on behalf of the national authority.

>Study audit means a comparison of the raw data and associated records with the final or interim report, in order to determine that the study was carried out in accordance with the study plan and standard operating procedures.

>To check means "to examine critically or to verify".

4.3: PROCEDURES

PRE-INSPECTION PROCEDURES

Prior to visiting a test facility, the inspector should become familiar with any existing information relevant to the subject of the inspection. Such information might include previous inspection reports, description of the facility, study plans, or study reports. The test facility to be inspected is usually notified in advance of the scheduled visit to ensure that the required personnel will be in attendance and that relevant records are available.

INSPECTION AND STUDY AUDIT PROCEDURES

The inspector should provide identification to authorised representatives of the test facility and should briefly discuss with them the purpose and nature of the inspection and/or study audit. The inspector should request access to any relevant documents or other information such as study plans, standard operating procedures, study reports, and raw data which are required to complete the inspection and/or study audit. A complete inspection should cover the following specific items; for a study audit, only appropriate items should be covered.

Test Facility Organisation and Personnel

Purpose: To check that the test facility organisation adheres to the OECD Principles of Good Laboratory Practice and to attempt to determine whether the test facility has an adequate number of qualified personnel.

(1) Check the organisational structure within the test facility, including identification and qualifications of individuals serving as Study Directors.

(2) Check that there are personnel training programmes including both on-the-job training and outside training.

(3) For studies (ongoing or completed) selected for inspection and/or audit:

(a) Obtain names, job descriptions, and summaries of training and experience for selected personnel engaged in the study(s) such as Study Directors and principal scientists;

(b) Attempt to determine whether selected individuals participating in the study had time to accomplish tasks specified by the study plan.

Quality Assurance Programme

Purpose: To check the mechanism by which the test facility management is assured that studies are conducted in accordance with OECD Principles of Good Laboratory Practice.

(1) Check that a Quality Assurance Programme is in operation.

(2) Interview the Quality Assurance Programme personnel to obtain information on how they schedule and conduct inspections and/or audits and to whom they report.

Facilities

Purpose: To check that the facilities are suitable so that the studies may be conducted in accordance with OECD Principles of Good Laboratory Practice.

(1) Check that the size and design of the test facility are applicable to the type of studies conducted therein.

(2) Check environmental control and monitoring procedures in important areas, such as animal rooms, test substance storage areas, laboratory areas.

(3) Check current cleanliness of the facilities and appropriate use of pest control procedures.

(4) Check that separation is maintained in rooms or areas where functions requiring separation are performed.

Apparatus, Materials, and Reagents

Purpose: To check that the test facility has suitably located apparatus in sufficient quantity and of adequate capacity to meet the requirements of the tests conducted and that materials and reagents are properly labelled, used, and stored.

(1) Check the general condition and cleanliness of apparatus.

(2) Check the records of apparatus operation, maintenance, standardisation, and calibration.

(3) Check that reagents are properly labelled and stored.

(4) Check that apparatus and materials used have not interfered with the test system.

Test Systems

Purpose: To check that test systems (animal, plant, microbial, cellular, sub-cellular, chemical or physical) are adequately accommodated and controlled.

Physical/Chemical

(1) Check as indicated in the section on apparatus, materials and reagents, above. Determine whether reference substances were used.

Biological

(1) Check that the test system is as specified in the study plan.

(2) Check the records of receipt, handling, housing, and care of test systems.

(3) Check that there are provisions for health evaluation of animal and plant test systems and their isolation, if necessary.

(4) Check that written records are kept of examination, quarantine, morbidity, mortality, and diagnosis and treatment of animal and plant test systems.

(5) Check that test systems are adequately identified.

(6) Check that the environment is as specified in the study plan or standard operating procedures such as housing, temperature, humidity, and light/dark cycles.

(7) Check that there are provisions for the appropriate disposal of the test system.

Test and Reference Substances

Purpose: To check procedures designed to ensure that the identity, quantity, and composition of test and reference substances administered to test systems are in accordance with the study plan.

(1) Check that there are procedures for receipt, handling, sampling, and storage of test and reference substances.

(2) Check that there are procedures for the determination of identity, purity, composition, stability, and prevention of contamination of test and reference substances, when applicable.

(3) Check the labelling of test and reference substances and that records are kept of their composition, characterisation, concentration, and stability, as applicable.

(4) Check that there are procedures for the determination of the homogeneity and stability of mixtures containing test or reference substances, when applicable.

(5) Check the labelling of mixtures containing the test or reference substances and that records are kept of homogeneity and stability, as applicable.

(6) Check that samples from each batch of test substances have been taken for analytical purposes and that they have been retained when the test is of a longer duration than four weeks.

Standard Operating Procedures

Purpose: To check that the test facility has written standard operating procedures that are related to the tests conducted.

(1) Check that each laboratory unit has immediately available the relevant written standard operating procedures.

(2) Check that standard operating procedures are available for, but not necessarily limited to, the following activities:

- (a) Receipt, identification, labelling, handling, sampling, and storage of test and reference substances;

- (b) Maintenance, cleaning, calibration of measuring apparatus and environmental control equipment, and preparation of reagents;

- (c) Recordkeeping, reporting, storage, and retrieval thereof;

- (d) Preparation and environmental control of areas containing the test system;

- (e) Receipt, transfer, location, characterisation, identification, and care of test systems;

- (f) Handling of the test systems before, during, and at the termination of the study;

- (g) Handling and disposal of test systems;

- (h) Use of pest control and cleaning agents;

- (i) Quality Assurance Programme operations.

(3) Check that any changes in standard operating procedures are authorised and dated.

Performance of the Study

Purpose: To check that the study plan and the study conduct are in accordance with the OECD Principles of Good Laboratory Practice.

(1) Check that written study plans exist and that they include the elements specified in Section 8.2 of the OECD Principles of Good Laboratory Practice.

(2) Check that information concerning the sponsor and the test facility, including signatures and dates of approval, are recorded.

(3) Check that amendments to study plans are signed and dated.

(4) Check that measurements, observations, and examinations are in accordance with the study plan and the standard operating procedures.

(5) Check that records of the raw data and any changes therein are made in accordance with the OECD Principles of Good Laboratory Practice.

(6) Audit the results presented in the reports of the study (interim or final) for consistency and completeness to determine that they correctly reflect the raw data.

(7) Check that any unforseen events recorded in the raw data are evaluated.

Reporting of Study Results

Purpose: To check that a final report has been prepared in accordance with the OECD Principles of Good Laboratory Practice.

(1) Check that when a final report is available, it is signed and dated by the responsible personnel.

(2) Check that amendments are made by the responsible personnel.

(3) Check that a quality assurance statement is included in the final report and that it is signed and dated.

(4) Check that the final report lists the location of all samples, specimens, and raw data.

Storage and Retention of Records and Materials

Purpose: To check that provision is made for the storage and retention of records and materials.

(1) Check that archives exist for the storage of study plans, raw data, final reports, samples, and specimens.

(2) Check the procedure for retrieval of archived materials.

(3) Check the procedure whereby access is limited to authorised personnel.

(4) Check that records and materials are retained for the required period of time, as appropriate.

COMPLETION OF THE INSPECTION

Consultation

The inspection visit ends with a consultation between the inspector and representatives of the test facility. The purpose of the consultation is to discuss the inspectional observations.

Inspection Report

The results of the inspection visit will take the form of a written report which will cover all aspects of the inspector's observations. The format of the report should parallel the sequence of items as found in the OECD Principles of Good Laboratory Practice and in this document.

ANNEX I

MANDATE OF THE GROUP OF EXPERTS

[C(78)127(Final) Appendix]

GOOD LABORATORY PRACTICE

1. Agreement between countries on basic standards to govern the acquisition of data respecting potential hazards to human health and the environment is an important step towards mutual acceptance of the data. It is proposed, therefore, that consideration be given to developing acceptable laboratory standards for the generation of laboratory data for regulatory purposes in OECD Member countries. In particular, attention should be given to:

    (i) specifying minimal acceptable laboratory practices governing tests to be specified by the current OECD Chemicals Testing Programme;

    (ii) record-keeping procedures, storage and accessibility for national control purposes of records and minimum time-period during which records and specimens must be retained after completion of a study;

    (iii) qualifications and experience required of personnel (professional and technical).

2. It should be recognised that the objective is to establish basic standards although emphasizing the desirability of encouraging the development and universal application of the highest possible standards of laboratory investigations.

3. In elaborating the basic laboratory practices, due cognizance should be given to existing authoritative douments on this subject. In particular, the forthcoming report from WHO on the principles for evaluating the toxicity of chemicals, and documents such as the United States Good

Laboratory Practices Regulations and the Compliance Programme pertaining to these Regulations, should be referred to.

4. It is also proposed that mutually acceptable means be developed to ensure that data in respect of hazards to human health and the environment is obtained in accordance with these basic standards. Bearing in mind the technical feasibility and practicality of implementing procedures that may be recommended, it is proposed to examine:

- (i) systems of accreditation and/or inspection of laboratories existing or proposed in each country and by international organisations;

- (ii) means of harmonizing such systems;

- (iii) certification (validation) procedures to ensure that acceptable laboratory practices have been employed in connection with any particular investigation;

- (iv) the need to establish exchange of information on national accreditation schemes (bearing in mind confidentiality aspects, appropriate time for exchange, etc,);

- (v) the need to make provision for approriate action against laboratories to meet acceptable standards.

ANNEX II

EXCERPTS FROM THE
DECISION OF THE OECD COUNCIL
CONCERNING THE MUTUAL ACCEPTANCE OF DATA
IN THE ASSESSMENT OF CHEMICALS

C(81)30 (Final), adopted on 12th May, 1981

THE COUNCIL

PART I

1. DECIDES that data generated in the testing of chemicals in an OECD Member country in accordance with OECD Test Guidelines and OECD Principles of Good Laboratory Practice shall be accepted in other Member countries for purposes of assessment and other uses relating to the protection of man and the environment.

2. DECIDES that for the purposes of this Decision and other Council actions the terms OECD Test Guidelines and OECD Principles of Good Laboratory Practice shall mean guidelines and principles adopted by the Council.

3. INSTRUCTS the Environment Committee to review action taken by Member countries in pursurance of this Decision and to report periodically thereon to the Council.

4. INSTRUCTS the Environment Committee to pursue a programme of work designed to facilitate implementation of this Decision with a view to establishing further agreement on assessment and control of chemicals within Member countries.

PART II

To implement the Decision set forth in Part I:

1. RECOMMENDS that Member countries, in the testing of chemicals, apply the OECD Test Guidelines and the OECD Principles of Good Laboratory Practice, set forth respectively in Annexes 1 and 2(*) which are integral parts of this text.

(*) OECD Guidelines for Testing of Chemicals" (1981 and continuing series). OECD: Paris and Chapter 2 of this volume respectively.

INSTRUCTS the Management Committee of the Special Programme on the Control of Chemicals in conjunction with the Chemicals Group of the Environment Committee to establish an updating mechanism to ensure that the aforementioned test guidelines are modified from time to time as required through the revision of existing Guidelines or the development of new Guidelines.

3. INSTRUCTS the Management Committee of the Special Programme on the Control of Chemicals to pursue its programme of work in such a manner as to facilitate internationally harmonized approaches to assuring compliance with the OECD Principles of Good Laboratory Practice and to report periodically thereon.

ANNEX III

EXPERT GROUP ON GOOD LABORATORY PRACTICE

LIST OF EXPERTS

AUSTRALIA	I. Bell, I. Carruthers, J.A. Gilmour, R.L. Norris, J.H. Whittem
AUSTRIA	H. Hofer
BELGIUM	L. Delcour-Firquet, J.-P. Tassignon
CANADA	D.L. Arnold, K.A. McCully, J.F. Riou, R. Sauriol, W.M.J. Strachan, R. Willis
DENMARK	P. Dalager, G. Lehmann-Nielsen
FRANCE	R. Cabridenc, J. Cordonnier, R. Glomot, R. Grech, P. Peirani, M.-P. Serre
GERMANY	N.C. Franklin, K.H. Leist, N. Lingk, D. Kayser, H. Schulze, A. Somogyi, B. Stalder
GREECE	V. Sotiriadou
ITALY	R. Amici, G. Della Porta, G. Falconi
JAPAN	H. Imura, T. Kariyone, K. Kobayashi, M. Kitano, M. Masuda, T. Omori, K. Saito, T. Yoshida
NETHERLANDS	D.M.M. Adema, M.J. van Logten
NEW ZEALAND	J.H. Garside
NORWAY	T. Sanner
SWEDEN	U. von Haartman
SWITZERLAND	W. Basler, H. Kelterborn

UNITED KINGDOM	M. Draper, F. Fairweather, F.J. Fielder, A.P. Fletcher, B.H. MacGibbon, C.R. Pearson, G.J. Turnbull, E.M.B. Smith, M. van der Heuvel
UNITED STATES	E.L. Brisson, D.D. McCollister, F.E. Freeburg, T.E. Hunt, P. Lepore, C.R. Morris (Chairman), D.M. Reisa, A.K. Stern, R.K. Tucker, L. Turner
COMMISSION OF THE EUROPEAN COMMUNITIES	R. Amavis, A. Berlin, R. Haigh, G. Mosselmans, W. Schäfer, Th. van der Venne
OECD	P.J. Crawford, B.O. Wagner
WHO (Observers)	H.L. Falk, G.J. Turnbull
ISO/CERTICO (Observers)	R.J. Amorosi, D. Volkmann F. Tricoche

OECD SALES AGENTS
DÉPOSITAIRES DES PUBLICATIONS DE L'OCDE

ARGENTINA – ARGENTINE
Carlos Hirsch S.R.L., Florida 165, 4° Piso (Galería Guemes)
1333 BUENOS AIRES, Tel. 33.1787.2391 y 30.7122
AUSTRALIA – AUSTRALIE
Australia and New Zealand Book Company Pty, Ltd.,
10 Aquatic Drive, Frenchs Forest, N.S.W. 2086
P.O. Box 459, BROOKVALE, N.S.W. 2100
AUSTRIA – AUTRICHE
OECD Publications and Information Center
4 Simrockstrasse 5300 BONN. Tel. (0228) 21.60.45
Local Agent/Agent local :
Gerold and Co., Graben 31, WIEN 1. Tel. 52.22.35
BELGIUM – BELGIQUE
LCLS
35, avenue de Stalingrad, 1000 BRUXELLES. Tel. 02.512.89.74
BRAZIL – BRÉSIL
Mestre Jou S.A., Rua Guaipa 518,
Caixa Postal 24090, 05089 SAO PAULO 10. Tel. 261.1920
Rua Senador Dantas 19 s/205-6, RIO DE JANEIRO GB.
Tel. 232.07.32
CANADA
Renouf Publishing Company Limited,
2182 St. Catherine Street West,
MONTRÉAL, Que. H3H 1M7. Tel. (514)937.3519
OTTAWA, Ont. K1P 5A6, 61 Sparks Street
DENMARK – DANEMARK
Munksgaard Export and Subscription Service
35, Nørre Søgade
DK 1370 KØBENHAVN K. Tel. +45.1.12.85.70
FINLAND – FINLANDE
Akateeminen Kirjakauppa
Keskuskatu 1, 00100 HELSINKI 10. Tel. 65.11.22
FRANCE
Bureau des Publications de l'OCDE,
2 rue André-Pascal, 75775 PARIS CEDEX 16. Tel. (1) 524.81.67
Principal correspondant :
13602 AIX-EN-PROVENCE : Librairie de l'Université.
Tel. 26.18.08
GERMANY – ALLEMAGNE
OECD Publications and Information Center
4 Simrockstrasse 5300 BONN Tel. (0228) 21.60.45
GREECE – GRÈCE
Librairie Kauffmann, 28 rue du Stade,
ATHÈNES 132. Tel. 322.21.60
HONG-KONG
Government Information Services,
Publications/Sales Section, Baskerville House,
2/F., 22 Ice House Street
ICELAND – ISLANDE
Snaebjörn Jönsson and Co., h.f.,
Hafnarstraeti 4 and 9, P.O.B. 1131, REYKJAVIK.
Tel. 13133/14281/11936
INDIA – INDE
Oxford Book and Stationery Co. :
NEW DELHI-1, Scindia House. Tel. 45896
CALCUTTA 700016, 17 Park Street. Tel. 240832
INDONESIA – INDONÉSIE
PDIN-LIPI, P.O. Box 3065/JKT., JAKARTA, Tel. 583467
IRELAND – IRLANDE
TDC Publishers – Library Suppliers
12 North Frederick Street, DUBLIN 1 Tel. 744835-749677
ITALY – ITALIE
Libreria Commissionaria Sansoni :
Via Lamarmora 45, 50121 FIRENZE. Tel. 579751/584468
Via Bartolini 29, 20155 MILANO. Tel. 365083
Sub-depositari :
Ugo Tassi
Via A. Farnese 28, 00192 ROMA. Tel. 310590
Editrice e Libreria Herder,
Piazza Montecitorio 120, 00186 ROMA. Tel. 6794628
Costantino Ercolano, Via Generale Orsini 46, 80132 NAPOLI. Tel. 405210
Libreria Hoepli, Via Hoepli 5, 20121 MILANO. Tel. 865446
Libreria Scientifica, Dott. Lucio de Biasio "Aeiou"
Via Meravigli 16, 20123 MILANO Tel. 807679
Libreria Zanichelli
Piazza Galvani 1/A, 40124 Bologna Tel. 237389
Libreria Lattes, Via Garibaldi 3, 10122 TORINO. Tel. 519274
La diffusione delle edizioni OCSE è inoltre assicurata dalle migliori librerie nelle città più importanti.
JAPAN – JAPON
OECD Publications and Information Center,
Landic Akasaka Bldg., 2-3-4 Akasaka,
Minato-ku, TOKYO 107 Tel. 586.2016
KOREA – CORÉE
Pan Korea Book Corporation,
P.O. Box n° 101 Kwangwhamun, SÉOUL. Tel. 72.7369

LEBANON – LIBAN
Documenta Scientifica/Redico,
Edison Building, Bliss Street, P.O. Box 5641, BEIRUT.
Tel. 354429 – 344425
MALAYSIA – MALAISIE
and/et SINGAPORE - SINGAPOUR
University of Malaya Co-operative Bookshop Ltd.
P.O. Box 1127, Jalan Pantai Baru
KUALA LUMPUR. Tel. 51425, 54058, 54361
THE NETHERLANDS – PAYS-BAS
Staatsuitgeverij
Verzendboekhandel Chr. Plantijnstraat 1
Postbus 20014
2500 EA S-GRAVENHAGE. Tel. nr. 070.789911
Voor bestellingen: Tel. 070.789208
NEW ZEALAND – NOUVELLE-ZÉLANDE
Publications Section,
Government Printing Office Bookshops:
AUCKLAND: Retail Bookshop: 25 Rutland Street,
Mail Orders: 85 Beach Road, Private Bag C.P.O.
HAMILTON: Retail Ward Street,
Mail Orders, P.O. Box 857
WELLINGTON: Retail: Mulgrave Street (Head Office),
Cubacade World Trade Centre
Mail Orders: Private Bag
CHRISTCHURCH: Retail: 159 Hereford Street,
Mail Orders: Private Bag
DUNEDIN: Retail: Princes Street
Mail Order: P.O. Box 1104
NORWAY – NORVÈGE
J.G. TANUM A/S Karl Johansgate 43
P.O. Box 1177 Sentrum OSLO 1. Tel. (02) 80.12.60
PAKISTAN
Mirza Book Agency, 65 Shahrah Quaid-E-Azam, LAHORE 3.
Tel. 66839
PHILIPPINES
National Book Store, Inc.
Library Services Division, P.O. Box 1934, MANILA.
Tel. Nos. 49.43.06 to 09, 40.53.45, 49.45.12
PORTUGAL
Livraria Portugal, Rua do Carmo 70-74,
1117 LISBOA CODEX. Tel. 360582/3
SPAIN – ESPAGNE
Mundi-Prensa Libros, S.A.
Castelló 37, Apartado 1223, MADRID-1. Tel. 275.46.55
Libreria Bosch, Ronda Universidad 11, BARCELONA 7.
Tel. 317.53.08, 317.53.58
SWEDEN – SUÈDE
AB CE Fritzes Kungl Hovbokhandel,
Box 16 356, S 103 27 STH, Regeringsgatan 12,
DS STOCKHOLM. Tel. 08/23.89.00
SWITZERLAND – SUISSE
OECD Publications and Information Center
4 Simrockstrasse 5300 BONN. Tel. (0228) 21.60.45
Local Agents/Agents locaux
Librairie Payot, 6 rue Grenus, 1211 GENÈVE 11. Tel. 022.31.89.50
TAIWAN – FORMOSE
Good Faith Worldwide Int'l Co., Ltd.
9th floor, No. 118, Sec. 2
Chung Hsiao E. Road
TAIPEI. Tel. 391.7396/391.7397
THAILAND – THAILANDE
Suksit Siam Co., Ltd., 1715 Rama IV Rd,
Samyan, BANGKOK 5. Tel. 2511630
TURKEY – TURQUIE
Kültur Yayinlari Is-Türk Ltd. Sti.
Atatürk Bulvari No : 77/B
KIZILAY/ANKARA. Tel. 17 02 66
Dolmabahce Cad. No : 29
BESIKTAS/ISTANBUL. Tel. 60 71 88
UNITED KINGDOM – ROYAUME-UNI
H.M. Stationery Office, P.O.B. 569,
LONDON SE1 9NH. Tel. 01.928.6977, Ext. 410 or
49 High Holborn, LONDON WC1V 6 HB (personal callers)
Branches at: EDINBURGH, BIRMINGHAM, BRISTOL,
MANCHESTER, BELFAST.
UNITED STATES OF AMERICA – ÉTATS-UNIS
OECD Publications and Information Center, Suite 1207,
1750 Pennsylvania Ave., N.W. WASHINGTON, D.C.20006 – 4582
Tel. (202) 724.1857
VENEZUELA
Libreria del Este, Avda. F. Miranda 52, Edificio Galipan,
CARACAS 106. Tel. 32.23.01/33.26.04/33.24.73
YUGOSLAVIA – YOUGOSLAVIE
Jugoslovenska Knjiga, Terazije 27, P.O.B. 36, BEOGRAD.
Tel. 621.992

Les commandes provenant de pays où l'OCDE n'a pas encore désigné de dépositaire peuvent être adressées à :
OCDE, Bureau des Publications, 2, rue André-Pascal, 75775 PARIS CEDEX 16.
Orders and inquiries from countries where sales agents have not yet been appointed may be sent to:
OECD, Publications Office, 2 rue André-Pascal, 75775 PARIS CEDEX 16.

65579-9-1982

OECD PUBLICATIONS, 2, rue André-Pascal, 75775 PARIS CEDEX 16 - No. 42353 1982
PRINTED IN FRANCE
(59 82 04 1) ISBN 92-64-12367-9